AF491704

GRAPH PAPER COMPOSITION NOTEBOOK QUAD RULE

This book belongs to:

____

If found please contact:

____

____

What do you think of this book?

We'd **LOVE** you to leave a review! We are just a tiny family business, and the

comments and feedback you share help us to create better books for you.

Thank you from us all at **The Careful Composition Company** - you are the best!

COPYRIGHT © 2020 - THE CAREFUL COMPOSITION COMPANY. ALL RIGHTS RESERVED.

www.ingramcontent.com/pod-product-compliance
Lightning Source LLC
Chambersburg PA
CBHW081340160726
48000CB00010B/3174